BEI GRIN MACHT SICH IHR WISSEN BEZAHLT

- Wir veröffentlichen Ihre Hausarbeit, Bachelor- und Masterarbeit

- Ihr eigenes eBook und Buch - weltweit in allen wichtigen Shops

- Verdienen Sie an jedem Verkauf

Jetzt bei www.GRIN.com hochladen und kostenlos publizieren

Impressum:

Copyright © 2018 GRIN Verlag
Druck und Bindung: Books on Demand GmbH, Norderstedt Germany
ISBN: 9783668761803

Dieses Buch bei GRIN:

https://www.grin.com/document/435052

Johann Gruber-Schmidt

Dibutyl Ether as Fuel in Aviation Drones

GRIN Verlag

Inhaltsverzeichnis

1. Introduction

Mobility is a very important property in trade and economy. Additional we have a high personal mobility and a mobility of loads transported from a point A to the next one B in a very short time. Using dimethyl ether as a substitute for fossil diesel in heavy truck has been proven by VOLVO since 2008 [4]. VOLVO applied dimethyl ether in heavy trucks running on the road over five years showing the benefits of dimethyl ether in the very high reduction of exhaust pollution emission [4[7[8]. But Dimethyl Ether is used in heavy duty trucks running on the road [8]. Trying to use dimethyl ether in Avionics Cause a lot of problems, therefore we have developed dimethyl ether further to dibutyl ether. Dibutyl ether as a result of the development [9]has a high application potential in avionics.

2. Kerosene versus Dibutyl Ether

Today one fuel often used in aviation in motors and gas turbines is kerosene fuel [6]. The properties of common kerosene [6] are given: ignition temperature 220°C, density 775 up to 820 kg/m³, boiling temperature 75°C up to 250°C, temperature class: T3, heat caloric value H_u= 34 MJ/L (800 kg/m³))= 9.5 kWH/L, viscosity v = 8.0 mm²/sec (T= -20°C) [7]. Kerosene is a fossil fuel.

Comparing kerosene with dibutyl ether can only be done on the thermodynamic data, because kerosene is often used in Otto engines and in combination with oil in gas turbines. Dibutyl ether is often used in diesel engines and in combination with additives in gas turbines.

The properties of common dibutyl ether [7[9]are given: ignition temperature 250°C, density 775 up to 820 kg/m³, boiling temperature 140°C up to 250°C, temperature class: T3, heat caloric value H_u= 43 MJ/L (800 kg/m³)) = 10.5 kWh/L, viscosity v = 8.0 mm²/sec (T= -20°C). Dibutyl ether is a renewable organic fuel. [9]

Comparing the main thermodynamic properties between both fuels we see that both fuels have nearly the same properties, and the possibility is given to substitute kerosene with dibutyl ether as fuel for gas turbines.

3. Dibutyl Ether Zero Power Generation

The pathway to dibutyl ether is running over the production of dimethyl ether. Dimethyl Ether is produced from synthetic gas, carbon monoxide (CO) and hydrogen (H_2), or carbon dioxide and hydrogen (H_2) with a catalyst under low pressure (pressure range $p = 30$ bar up to 100 bar) at a temperature of $T = 200°C$ up to $300°C$. In most cases dimethyl ether is produced over the intermediate product methanol (CH_3OH), and the Dehydration of methanol to dimethyl ether. [8]

Chemical reactions for the pathway over carbon monoxide and hydrogen:

$$CO + 2H_2 \longrightarrow CH_3OH + Q\,(-\ 95.4\ kJ/mol)$$
$$2CH_3OH \longrightarrow CH_3OCH_3 + H_2O + Q\,(-\ 18.2\ kJ/mol)$$

[E 1]

Chemical reactions for the pathway over carbon dioxide and hydrogen:

$$CO_2 + 3H_2 \longrightarrow CH_3OH + H_2O + Q\,(-\ 10.4\ kJ/mol)$$
$$2CH_3OH \longrightarrow CH_3OCH_3 + H_2O + Q\,(-\ 18.2\ kJ/mol)$$

[E 2]

With Dimethyl ether using hydrogenation and carbonylation supported by catalysts Dimethyl ether is converted to methanol and ethanol. Methanol is recycled and ethanol is used to be converted to butanol over a catalyst. Butanol is reduced by dehydration supported by a catalyst to dibutyl ether. The chemical reactions are listed up:

$$CH_3OCH_3 + CO + 2H_2 \longrightarrow CH_3OH + CH_3CH_2OH + Q\,(-\ 145.6\ kJ/mol)$$
$$2CH_3CH_2OH \longrightarrow CH_3(CH_2)_3OH + H_2O + Q\,(-\ 54.1\ kJ/mol)$$
$$2CH_3(CH_2)_3OH \longrightarrow CH_3(CH2)_3O(CH_2)_3CH_3 + H_2O + Q\,(-\ 68.0\ kJ/mol)$$

[E 3]

Dibutyl ether can be stored in a tank at environment pressure. The evaporation temperature is given with a temperature T ~ 150°C. Dibutyl ether as a fuel can be converted to electric energy and thermal heat with the following process: The fuel dibutyl ether is heated up and filled in the combustion swing chamber. In the combustion chamber dibutyl ether is converted with oxygen to carbon dioxide (CO_2) and hydrogen(H_2):

$$C_8H_{18}O + O_2 \longrightarrow 8CO_2 + 9H_2O + Q(-\ 5347.0\ kJ/mol) \qquad [E\ 4]$$

The hot exhaust gas has a temperature of 1600°C up to 2000°C. The highly exergetic heat is converted with a closed magneto hydrodynamic (MHD) generator system to electricity and heat. In the application the number of MHD-generators is at least one and maximum four. The liquid carbon dioxide (CO_2) and water (H_2O) are stored in tanks. The operating pressure of the exhaust system is about p = 70 bar. Water steam condensation takes place at a temperature of T=263°C and carbon dioxide condensation takes place at a temperature of 25°C. [1[2[3]

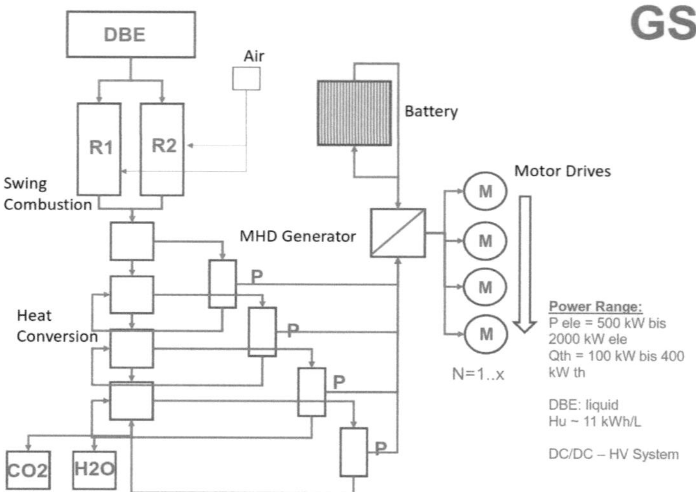

Figure 1: Zero Emission dibutyl ether with Plasma Generator based on the MHD technology, storing carbon dioxide and water in liquid phase in tanks. (Source: Johann Gruber-Schmidt, 2018)

In the figure [1] a simple power cycle based on dibutyl ether is shown. The fuel dibutyl ether is led into a swing combustion chamber, in which the fuel with oxygen is oxidized to carbon dioxide and steam. The hot exhaust has temperature between T=1600°C up to T=2000°C. If we have an operating pressure p~ 70 bar, steam condensing can be realized at a temperature T~ 260°C. At a pressure of p~ 70 bar carbon dioxide condensing takes place at a temperature T~ 25°C.

The heat is converted in a small closed magnetohydrodynamic cycle. The generator is operating based on the Hall effect. The cascade of several MHD generators enable to increase the electric efficiency. The generator produces electric power under constant voltage. (DC). The power module includes also a battery system, to stabilize the controlling and measurement system. The electric power is used for electric power driving motors.

The condensed water is collected in a tank, the condensed carbon dioxide is collected in pressurized tanks (p~ 70 bar).

The oxygen for oxidation of dibutyl ether is take from the air. The air is preheated and led into the combustion reactor. Using two reactors lead to a combustion swing system is so very close to the continuous process.

4. Drones

One application of the Zero Emission Fuel is to use dibutyl ether as a fuel for drones. Drones are well-known and the next step is to enlarge the drones for passenger transportation and container transportation. The drone is an element of the three-dimensional space and enlarges the mobility from the plane (surface) up to the near surface earth space. This is a necessary step to increase the mobility in transportation and to stop wasting time, if the cars and truck are waiting in an unexpected traffic hold up. [13]

Drones have the advantage to act individual and free, so the flexibility of transportation and mobility is increased dramatically. Inside the drone, the passenger can watch and check the transportation, if necessary they can start emergency operation.

The propulsion for drones in this short article is analyzed to be done by propeller propulsion or electric plasma propulsion [11]. The thrust F[N] for a drone must be greater equal the own weight and the load to lift off, the thrust T[N] for moving forward is defined by travelling velocity u_T [m/sec]

$$F = (m_0 + m_L)g = m_{Air}(u_e - u_1)$$
$$T = (m_0 + m_L)u_T = m_{Air}(u_e - u_0)$$

[E 5]

The equation shows the equilibrium of the weight F[N] including the mass of the drone and the load and the thrust from the propulsion engines. The velocities of the flowing air mass (m_{Air}) are the entrance velocity in the propulsion engine and the exhaust air velocity u_1. For the travelling velocity u_T, we have the equilibrium force T[N]. The lifting of the drone and the travelling velocity of the drone are weak coupled.

The needed power P_1 [W] for the thrust F[N] and P_T [W] for the thrust T[N] are given by the velocity u_1 and u_T

$$P_1 = Fu_1 \qquad P_T = Tu_T \qquad P = P_1 + P_T \qquad\qquad [E\ 6]$$

The propulsion efficiency for lifting η_L and for travelling η_T is given by the approximate simple formula

$$h_{pL} = \frac{2}{1 + \dfrac{u_e}{u_1}} \qquad h_{pT} = \frac{2}{1 + \dfrac{u_e}{u_T}} \qquad\qquad [E\ 7]$$

The overall efficiency is given by the power needed to lift of the weight and to move forward in relation to the power of the needed mass fuel m_F [kg/sec] and the heat caloric value H_u [kJ/kg].

$$h = \frac{Tu_T + Fu_1}{m_F H_u} \qquad\qquad [E\ 8]$$

A drone does not have wings to fly and to start smooth form the ground. A drone has very simple operating possibilities: lift of from the ground, stabilizing in a certain travelling height from the ground, travelling along a defined way from point A to point B. This is shown by the weak coupling of lifting and travelling a drone.

The travelling speed is assumed to be about 100 km/h ~ 27 m/sec ~ u_e ~ u_T. This velocity is small compared with the exhaust velocity u_1~ 50 up to 100 m/sec of a propeller propulsion engine and u_1 ~ 100 up to 250 m/sec of a electric air breathing plasma propulsion [11] . According to the simple formulas above the most of the needed power is used for holding the drone in a definite height and small amount is used for travelling with speed.

5. Drones with Air propeller Propulsion

On possibility to realize the propulsion is using the well-known propeller engines, with two propeller and at least three blades for each propeller, driven by an electric motor. [10]

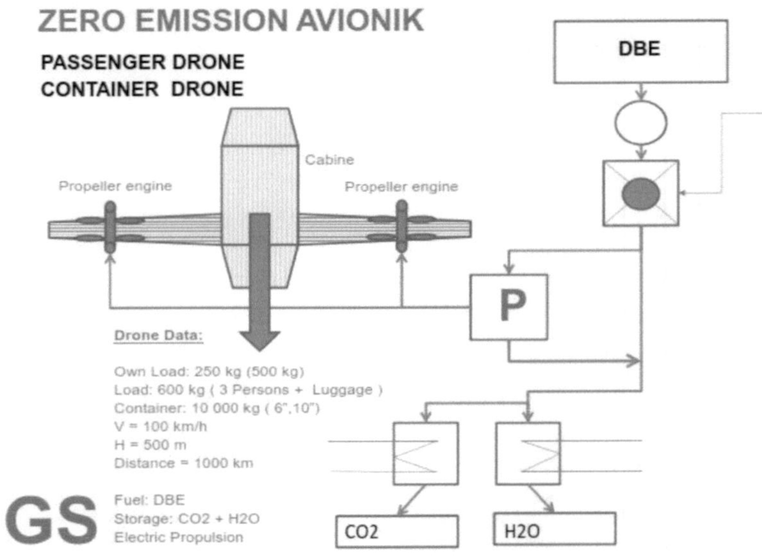

Figure 2: The Zero Emission Avionic based on the dibutyl ether fuel is realized by collecting the carbon dioxide and the water in tanks and the heat is converted to electric energy and thermal heat. The electric energy is used for the electric propulsion engines. (Source: Johann Gruber-Schmidt, 2018)

There are two limitations: the efficiency of the propeller itself converting the mechanical energy into thrust, and the second limitation is the electric motor, with the power, speed and moment, electric power P ~ 100 kW, speed n ~ 6000 (1/min) (rpm). The diameter of the propeller is given from d ~ 0.7 m up to 2.0 m, the number of blades at least N=3 up to N=9. The main disadvantage is the noise resulting from the propeller blade tips if they reach the sound velocity, even if the propeller is running in a casing.[10]

6. Drones with Air Breathing electric Propulsion

As we have shown in the last passage using propeller propulsion is limited by the noise resulting from the pressure waves from the propeller and in the speed of the propeller itself (tip velocity ~ 1.0 Mach sound speed). Another possibility is the air breathing electric propulsion. [11]

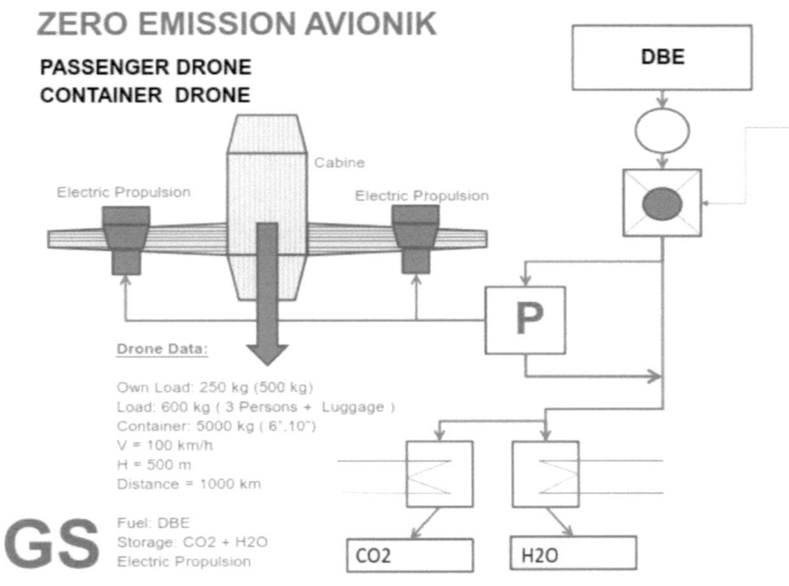

Figure 3: The Zero Emission Avionic based on the dibutyl ether fuel is realized by collecting the carbon dioxide and the water in tanks and the heat is converted to electric energy and thermal heat. The electric energy is used for the electric propulsion engines. [2[3] (Source: Johann Gruber-Schmidt, 2018)

Under electric air breathing propulsion we understand, air will be taken at inlet mass flow, compressed up to the pressure p_1 ($p_1 > p_0$), heated up, under very high magnetic fields the air will be ionized, and let into a magnetic nozzle, which consist of a cathode and anode. Such propulsion engines have the possibility to increase the outlet velocity of the nozzle, to increase the impulse and therefore to increase the thrust. The acoustic noise is much more less, we speak therefore from a quiet propulsion engine. Such strong engines enable to build up transportation drones, to transport heavy loads up to 10 t (100 kN) over long distances.

7. Personal Transport – Limitation

The limitation of passenger drones with air breathing electric propulsion is given by the ionization of the air. The property of Zero Emission has no limitation, carbon dioxide and water are collected in tanks, which are drained during the filling up of the fuel tanks.

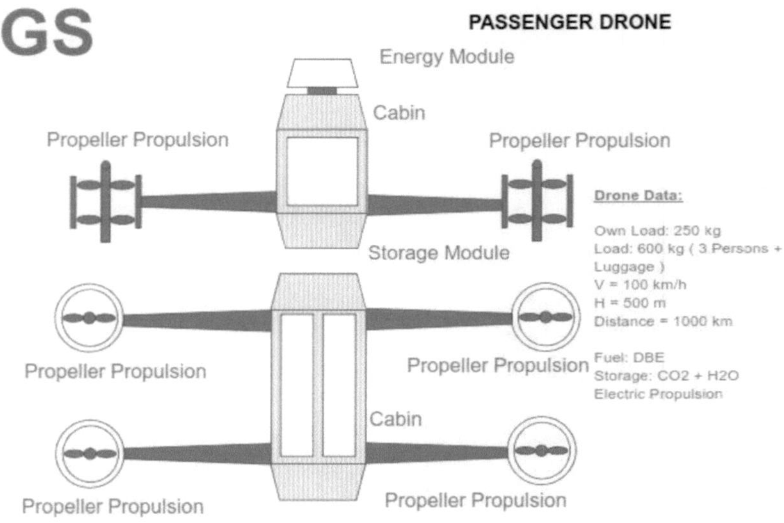

[11]

Figure 4: Personal Drones for transportation of passengers, operating in a maximum height of 500 m, operating in a distance range of 1000 km, based on the fuel dibutyl ether and Zero Emission. (Source: Johann Gruber-Schmidt, 2018)

The Number of propulsion engines is limited up to eight engines, each engine with a power of ~ P =25 kW ele, so that the maximum power needed is given with P=100 kWh ele. With an electric efficiency of η_{ele}~ 80% and η_{th} ~ 15% and 5% losses this leads to one tank with a tank volume of 600 liters, so that 500 L dibutyl ether is available. Additional we have one tank with 500 L for water and one tank with a volume each of 500 L for liquid carbon dioxide. Liquid carbon dioxide is stored at a pressure of p=70 bar. The heat generated by the conversion process is needed to heat the Energy module, to prevent the module of freezing and to heat up the fuel up to a temperature of T = 150°C to evaporate the liquid fuel to enter the surface combustion swing chambers as gaseous fuel. Additional we need the heat to provide a climate system for the passengers. [12[13]

The transportation drones are provided with a GPS system, and is running on a definite route between to defined points, the starting point A, and end point B. The controlling system of the drone is checked on the monitors of the base station. If failures are detected the drone can land safe. If other drones are crossing the path, the drone can go around. The controlling system is fail save system proven by the EN 62 061, EN 62 508.

8. Load Transport – Limitation

The limitation of container drones with air breathing electric propulsion is given by the ionization of the air. The property of Zero Emission has no limitation, carbon dioxide and water are collected in tanks, which are drained during the filling up of the fuel tanks.

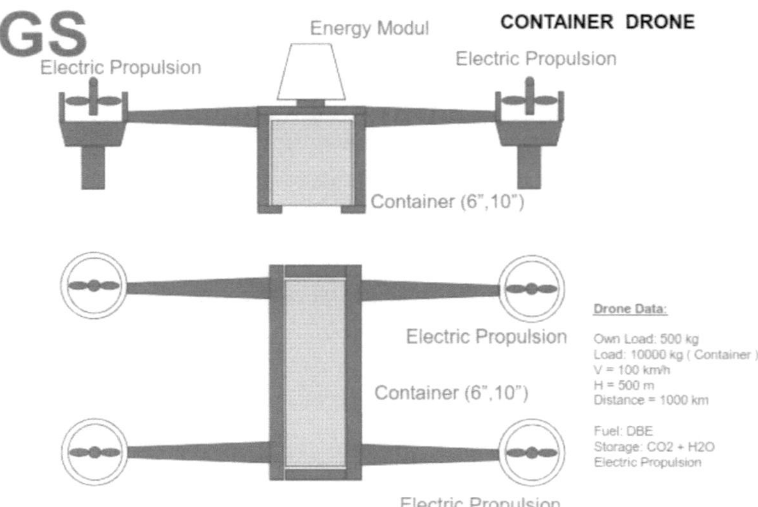

Figure 5: Container Drones for transportation of heavy loads, operating in a maximum height of 500 m, operating in a distance range of 1000 km, based on the fuel dibutyl ether and Zero Emission. (Source: Johann Gruber-Schmidt, 2018)

The Number of propulsion engines is limited up to eight engines, each engine with a power of ~ P =100 kW ele, so that the maximum power needed is given with P=800 kWh ele. With an electric efficiency of η_{ele}~ 80% and η_{th} ~ 15% and 5% losses this leads to two tanks with a tank volume of 600 liters and 1200 Liters, so that 1000 L Dibutyl Ether is available. Additional we have one tank with 1000 L for water and two Tanks with a volume each of 1000 L for liquid carbon dioxide. Liquid carbon dioxide is stored at a pressure of p=70 bar. The heat generated by the conversion process is needed to heat the Energy module, to prevent the module of freezing and to heat up the fuel up to a temperature of T = 150°C to evaporate the liquid fuel to enter the surface combustion swing chambers as gaseous fuel. [12]

The transportation drones are provided with a GPS system, and is running on a definite route between to defined points, the starting point A, and end point B. The controlling system of the drone is checked on the monitors of the base station. If failures are detected the drone can land safe. If other drones are crossing the path, the drone can go around. The controlling system is fail save system proven by the EN 62 061, EN 62 508.

9. Conclusion

A new generation of drones enables to transport passengers and container of short distances. Drones increase the mobility and flexibility in transportation of passengers and containers. The main property of all drone application is Zero Emission. Zero Emission is realized by storing of liquid carbon dioxide and liquid water. The propulsion engine is for small loads an electric propeller propulsion and for lager loads we use air breathing electric propulsion engines.

Zero Emission drones for passenger and container transport are a possible application of the renewable fuel dibutyl ether. [5]

Drawing and Pictures

10. References

[1]   Paul Connett, The zero Emission Solution, 2013, Chelsa Green Publ.

[2]   Takeshi Yao, Towards CO2 Zero Emission Systems, 2010, Springer Publ. York

[3]   E. Yantowsky et. Al., Zero Emission Power Cycles, 2009, CRC Press

[4]   VOLVO, Bio DME Heavy Truck, 2015, Company Flyer

[5]   Moazzem Hossaim et.al., Pathways towards a sustainable Economy, 2018, Springer Publ.

[6]   Safety data Sheet – Jet A 1 Fuel, ENI; 2018

[7]   California Dimethyl Ether Multimedia Evaluation TIER I, 2015

[8]   J.P. Szybist, Emissions and Performance Benchmarking of a Prototype Dimethyl Ether Fueled Heavy-Duty Truck, 2014, ORNL- TM 2014 – 59

[9]   J. Gruber-Schmidt, Dimethyl Ether and Dibutyl Ether produced from Biogas and Biomass, 2018, Presentation at Hagenbrunn/Austria

[10]   B.W. Jr. McCormik, Propeller Aerodynamics, 1956, Dover Publ.

[11]   T. Schönherr, Air breathing Electric propulsion, 2015, Univ. of. Tokyo

[12]   Dan M. Goebel, Fundamentals of electric propulsion, 2008, Wiley & Sons Publ.

[13]   Robert G. Jahn, Physics of electric propulsion, 1968, McGraw Hill Publ.

BEI GRIN MACHT SICH IHR WISSEN BEZAHLT

- Wir veröffentlichen Ihre Hausarbeit,
  Bachelor- und Masterarbeit

- Ihr eigenes eBook und Buch -
  weltweit in allen wichtigen Shops

- Verdienen Sie an jedem Verkauf

Jetzt bei www.GRIN.com hochladen
und kostenlos publizieren